the God Seed

God, He Who Created Hearts and Changes Minds

the God Seed

God, He Who Created Hearts and Changes Minds

K. O. Johnson-Luther, DTh

the God Seed—God, He Who Created Hearts and Changes Minds

Edited by: Jenny Margotta
            editorjennymargotta@mail.com

Cover by: Elissa J. Spaeth
Illustrations by: Kevin J. Collins

Printed in the United States

Second Edition: June 2022

Author

A native of New Jersey, K.O. Johnson-Luther received a Doctor of Theology in 2015, presenting a thesis during the program titled, *God, the One who Created Hearts and Changes Minds*. The knowledge discovered during this time of study and research so impacted her life and thinking that the endeavor to share this revealed truth was paramount.

Dr. Johnson-Luther has always been a lover of words and a people watcher. Even etymology was chosen as an elective in high school. Coupled with the desire to know and understand people, what causes us to behave so uniquely and the conveyance of our reasoning, there is a constant quest for awareness. She has always believed that we are agents of our minds, and the power of knowledge breeds the optimal ability to garner thoughts that create a life of achievement.

Dedication

To my children, Teneia and Bryan, grandchildren, Malachi and Levia, my brother, Eric, and those unnamed of generations to come.

To my parents, Maudie M. Brown and Morrise C. Johnson, because of the things you have deposited in my life, the views you have shared, and the lives you have lived, I have been positioned to take life seriously and contemplate it meticulously.

My prayer to God is that this will be a writing that shall effect change for generations, both in my lineage and those of the nation.

Acknowledgments

From inception and throughout the process of writing this book, great support has been provided to me from family and friends. Thank you.

I attribute the completion of this undertaking to the aptitudes afforded me from God, who has given me both the will and the ability to do all things. For that, I give Him all glory.

Foreword

I have always been fascinated with the intricacies and science of the human body the brain and how it relates to spirituality. When I read the prologue to this book, I was immediately intrigued. Like many people throughout the ages, as a young adult I wrestled with questions about the existence of an all-knowing God, my purpose, and why bad things happen. After becoming a Christian, I looked for ways to reconcile my questions with my love of science and knowledge.

The God Seed marries science, philosophy, and spirituality in an intelligent, well-researched format while remaining relatable and easy to understand.

Author, K. O. Johnson-Luther pursues truth in both science and religious faith, showing that good Bible study and good science do not conflict but, in fact, help us to see that God is loving and desires an intimate mind, heart, and soul connection with each human being.

~~Beckie Lindsey,
author of the award-winning
Beauties from Ashes series
and editor of *SoCal Christian Voice*

Notes from the Author

Graphics are for pictorial purposes only.

The King James version of the Hebrew-Greek Key Word Study Bible is used for scripture references, unless otherwise noted.

The reference to man is synonymous with woman, unless specified otherwise by context. Man is also used to reference humanity.

At first mention of words or phrases not commonly used or seen, definitions or explanations shall be provided as footnotes—designated 1, 2, 3 etc.—at the bottom of the page on which they appear.

Endnotes—designated a, b, c, etc.—appear at the end of the book.

All definitions have been obtained from the Merriam Webster dictionary, unless otherwise noted.

Reference to "the Word" is tantamount to the Bible. Scripture quotes will be *italicized*.

At points in the writing where characteristics, similitudes, or prototypes of God are revealed, they will be shown in a textbox titled "Footprint."

Table of Contents

Prologue 1

The Point of It All 3

The Origin of Man 6

Mental Capacity (Background) 16

The Brain—How It is Put Together 17

The Brain—How Its Parts Function 27

Consciousness and Memory 40

The Process of Memory Creation 49

The Plan and Science 63

The Heart and Mind 69

Renewing the Mind 80

Conclusion 88

Epilogue 90

Bibliography 93

Endnotes 98

There is a clue, a clue to meaning in life, and that clue comes in relationships. And the ultimate relationship is only when you find that relationship with God himself.

Ravi Zacharias

(1946-2020)

Prologue

This book, which began as a quest to show the connection between the mind, God, and the redemption of man, has also proven to be an expedition into the furtherance of the knowledge and understanding of God as Creator.

The creation, depicted in the book of Genesis, is just a surface representation of what He really created. It speaks little to the intricacies of what exists. Although *the God Seed* shall only speak to a minute aspect of the mind and its connection to its Creator, it is hoped that it will spark in the reader a desire to seek God for deeper revelation of what He has given.

Understanding the workings and abilities of the mind will enhance your perception of God. This knowledge will increase your faith in the ability you have, as the created being of God, to produce a life of fruitfulness—in whatever area you have been destined to thrive.

The following question is often asked:

Why didn't God create the perfect being, one that would follow Him without question or detour?

God created man and gave him the power of choice because He wanted to allow man to *choose* to love and follow Him. In His foreknowledge He left an eternal connection to Himself within us, which I call *the God Seed*. This was done so that, on the day we realize our desire for God—to accept Him as Lord and Savior—there would be a neurological connection in support of that appetency.

It is the desire of God that we would all endeavor to have the same mind that was in Christ Jesus, hence, the location of the remnant God seed in the mind of man.

The Point of It All

The late Arthur Fletcher, once head of the United Negro College Fund, coined the phrase, "A mind is a terrible thing to waste." The intricacies of the mind are utterly astonishing. Such a small member of the body, yet it wields such great power. As James 3:5 states, *Even so the tongue is a little member, and boasteth great things. Behold, how great a matter a little fire kindleth!*

This passage of scripture can be compared to the navigation of a great ship. The equipment used to steer the vessel, in size comparison to the ship, is considered small. The mind, a three-pound mass of tissue, neurons, and chemicals, can control not only bodies weighing an average of 65 times its weight,[a] but orchestrate the strategies of the world.

If viewed from a philosophical perspective, the creation of the world can teach science to scientific minds. After all, philosophy is the answering of questions that plague the minds of scientists. Embracing creationism uncovers answers that scientists have been seeking for decades. Leaders in science should consider the validity of the Bible's depiction of creation and the intricacies of the creation of man, particularly of his mind. In doing so they would discover the greatness of God, the

Creator. For an example, Psalm 19:1-6 of the Message Bible affirms that the skies show evidence of creation:

> *God's glory is on tour in the skies, God-craft on exhibit across the horizon. Madame Day holds classes every morning, Professor Night lectures each evening.*
>
> *Their words aren't heard, their voices aren't recorded, But their silence fills the earth: unspoken truth is spoken everywhere.*
>
> *God makes a huge dome for the sun – a superdome! The morning sun's a new husband leaping from his honeymoon bed, the daybreaking sun an athlete racing to the tape.*
>
> *That's how God's Word vaults across the skies from sunrise to sunset, melting ice, scorching deserts, warming hearts of faith.*

By denying the authenticity of creation, man has unknowingly been beguiled by the influences of the world in which he resides. The things or revelations of God are hidden purposely from those who refuse to accept Him as the Divine Creator of the world.[b] This has given man the proclivity of being a hater, faithless, and jealous.[c] The characteristics associated with this state of being are in total opposition to those desired for him by his Creator.

The denial has also procured a life for him that was never in the plan of his God. The Lord's greatest desire for us and Himself is a relationship that demonstrates mutual love. In His love for humanity, God has provided a means for us to return to Him through the shed blood of Jesus. *the God Seed* is a physical memory left of that relationship and mutual love.

This return requires the conversion of man's heart and the renewing of his mind. With the understanding that the body and its components were created and designed to depend upon one another, let us take an in-depth look at the mind and its original state. We will discover how the mind functions, how it can be renewed, the importance of its renewal, and the importance of the heart's role.

The Origin of Man

The God of creation describes man as a three-part living being created in His image and likeness. Man has a spirit, soul, and body. Isaiah 42:5 records this concerning creation and man:

- He created the heavens,
- He stretched them out,
- He spread out the earth and sustains the whole of it,
- He gave breath and spirit to them that walk upon it,
- There is a coming Savior.

Ephesians 4:10 declares that man is the workmanship of God. The Bible further confirms that man can think and live in oneness with his Creator. This oneness can be achieved by permitting the mind of Christ to be his. This can be accomplished by allowing his thinking and choices to align with the Word of God. Philippians 2:5 admonishes us to let the mind of Christ be ours, to allow His way of behavior to be ours. In Matthew 19:26, Jesus assures us that with Him, all things are possible.

God's plan for man was simply that he would be fruitful, multiply, replenish, and subdue the earth. He would have dominion over the fish of the sea, fowls of the air, cattle, and every living thing that moved upon the earth. He was placed in the Garden of Eden as a caretaker. All that man needed was provided by his Creator: *... and the Lord God planted a garden eastward in Eden; and there He put the man whom He had formed; and out of the ground made the Lord God to grow every tree that is pleasant to the sight and good for food ... and a river went out of Eden to water the garden ...[a]*

The text mentions only the provisions made for the sustaining of man's physical body because the spirit (heart) and soul (mind) of man existed naturally in unison with his Creator. Man, therefore, had no need for anything additional. Adam's (man, in the collective sense) only responsibility was simply to exist and adhere to the covenant between himself and his Creator.

Man is asked to totally rely upon God for direction and provision. This is not to imply that one is not required to put forth any effort in life, but in doing so, they must believe that all they endeavor to do will thrive. It is yet the desire of God to be the provider of our every need. Life of quality consists of more than money, popularity, success, and things. The Amplified Bible states:

> *Do not seek [by meditating and reasoning to inquire into] what you are to eat and what you are to drink; nor be of anxious (troubled) mind [unsettled, excited, worried, and in suspense]; for all the world*

is [greedily] seeking these things, and your Father knows that you need them (life-sustaining requirements, author emphasis added). *Only aim and strive for and seek His kingdom, and all these things shall be supplied to you. Do not be seized with alarm and struck with fear, little flock, for it is your Father's good pleasure to give you the kingdom!*[b](Greek meaning for the kingdom: royalty, rule, author added.)

While a child is under the care of its parents, the child need only be and adhere to the instructions of its parents, for it is the responsibility of the parents to supply the needs of the child. In Malachi 3:6 the Lord declares, *I am the Lord and I change not*. His decisions, plans, and desires for the well-being of man are the same as they were at creation.

Of all that Adam was given to follow, the most important was to be fruitful and increase the population of the earth. By doing so, he would eternalize and maintain God's kingdom agenda, which was and is for us to live in peace and prosperity. This has always been the Creator's desire. God wanted His image multiplied, an image that would perpetuate His constant love and desire upon the earth. He created us in His image and likeness because he wanted the inhabitants of the earth to live in harmony with one another and Himself.

An analogy of this type of image can be compared to a neighborhood community that shares like-minded respect for one another, each other's property, and children. Of course, He was not looking for, nor did He create, robots. This is evident in man's decision to do the one thing he was not to do: eat of the tree of the knowledge of good and evil. Even amid disobedience, God left man with the ability and presence of mind to be able to return to His original plan of emancipation.

When created, the trees were provided seed to multiply after their own kind; so, the God Seed is within man. This seed is one filled with the characteristics of God: loving, faithful, and patient.[c]

Now that man finds himself separated from God, how does he duplicate these characteristics and implement the lifestyle that exhibits the character of God? The answer to these questions can be acquired through the application of the principles of the Bible. The God-given characteristics of *the God Seed*, which lay dormant in the mind, are resuscitated by the activity of the Word of God in our lives.

The Word says, in Colossians 3:16 and Philippians 2:12-13, that if we allow the Word of Christ to dwell in us richly in all wisdom to the point that we are working out salvation with fear and trembling, God shall provide the will and ability to do of His good pleasure, which will duplicate His character in us. The Amplified Version of the Bible conveys these two scriptures in this way:

> *Let the word [spoken by] Christ (the Messiah) have its home [in your heart and mind] and dwell in you in [all its] richness,*

to the point that we

> *may (cultivate, carry out the goal, and fully complete) our own salvation with reverence and awe and trembling (self-distrust) with serious caution, tenderness of conscience, watchfulness against temptation, timidly shrinking from whatever might offend God and discredit the name of Christ. [Not in your own strength] for it is God Who is all the while effectually at work in us [energizing and creating in us the power and desire], both to will and to work for His good pleasure, satisfaction, and delight.*

The Greek meaning of the phrase *good pleasure* is "to accomplish or to perform." He will show us how and give us the capacity to live and remain in His will. His good pleasure for those who would believe in Him is to live full, joyful, peaceful, and prosperous lives. Romans 12:2b states that the believer can present himself converted from the world by the renewing of his mind. God says through John the Apostle in Third John 1:2, *Beloved, I wish above all things that thou mayest prosper and be in health, even as thy soul prospereth.*

It remains the desire of God for humanity to prosper, but there are stipulations. Since the creating of man and the making of woman, there has been a mandate from God for obedience. Society has a negative perspective on obedience. There has been a stigma associated with this word since the sin in the Garden. Obedience is meant to be a lifestyle and means by which the created displays his love and adoration for his Creator. In a relationship with Christ, man can share the original intent and purpose of obedience. Obedience from the perspective of God simply means that the creation reciprocates the love that the Creator has for His creation—us.

When the Lord stated to Israel in First Samuel 15:22, *it is better to obey than to sacrifice and to harken than the fat of rams,* he was simply saying that to love Me is better than your offerings of repentance. Your love for Me is all you need to walk in the ways destined for a life full of love, joy, and peace.

As God is righteous and just, it would be outside His character to require of man anything that he is incapable of achieving. The difficulty in man's obedience lies only in the limitations he places upon himself. He must only come to realize that he, as the being created in the image of God, has been fearfully and wonderfully made[1] with an inborn, distinctive characteristic and ability within him to achieve anything.

Prior to the violation of God's directive in the Garden of Eden, man was unacquainted with sin. He stood full of life and in total innocence before God, with the characteristics of God actively at work within him. Conversely, after the transgression of eating from the tree of the knowledge of good and evil, man, now exposed to sin, must negotiate between good and evil, right, and wrong, spirit and flesh, life, and death.

God, to extend grace to man during the interim of the birth, death, and the resurrection of Jesus, provided the Commandments or the Law to help man recognize his sinful nature and arduous plight. With the Law, God also provided a means for repentance, but the sinful nature of the heart of man, and his unregenerate[2] mind were always in control, inhibiting his ability to

---

1 Psalm 139:14 – *I will praise thee; for I am fearfully and wonderfully made: marvelous are thy works; and that my soul knoweth right well.*
2 Unregenerate, not reformed; unreconstructed.

perform what was right. Evil was continually present in the fallen heart and mind of man.

The birth and life of Jesus provided a rebirth of the man God created in the Garden. The first Adam indeed carried all the characteristics of his Creator. He had the ability to live as his Creator purposed, but this capacity was degraded through disobedience. God, in His love for His creation, provided Christ to offer a way to restore what was lost, calling Christ the second Adam. Through the second Adam, God offers man an example of a lifestyle that duplicates His characteristics. His life exemplified characteristics of purity of the heart and mind.

When God formed man, he was spiritually, naturally, and physically designed to do all God predetermined. We are created in the image and likeness of God and need only exercise this inherent potential of power. The God breathed Spirit; *Ruach*[3], given to man, empowered him with authority and dominion. During this time, the spirit of man and the Spirit of God functioned together as one.

It was through the Spirit of God that the world was spoken into existence. Man has within him authority in the spoken word. This

---

3 Ruach—the human spirit as breathed into man by God. Eventually, ruach came to mean the entire, immaterial consciousness of man.

authority of the spoken word is greatly enhanced when the Spirit of God resides synonymously with the spirit of man.

The Amplified Bible, in I Corinthians 6:17, makes this declaration concerning this co-habitation of the Spirit of God and that of man: *But the person who unites to the Lord becomes one spirit with Him.* When the Spirit of God governs the spirit of man, all the privileges planned for man in the Garden become available.

Genesis 1:26 states: *And God said, Let Us make man in Our image, after Our likeness….* In this verse the Hebrew definition of God, Elohiym,[d] renders the use of the noun God as plural, referring to the Father, Son, and Holy Spirit—the tri-part personality of the Godhead.

It is important now to understand the diversity between the definitions of image and likeness. These words have roots in the Hebrew language. As defined by *Strong's Exhaustive Concordance of the Bible,*[e]likeness, *damah,* and *demuwth* mean to compare; by implication to resemble, liken, model, shape. As image, *tselem* means to shade; a phantom (man's spirit), i.e. illusion; hence, a representative figure of the Spirit of God. Man has been formed with both the characteristic resemblance and spiritual authority of God.

In his book *Release of the Spirit*[f], Watchman Nee[g] defines tri-part as the inner man, spirit (heart); the outer man, soul (mind); and the outermost man, body. God intended for man's spirit or heart to be His dwelling place. Therefore, the Holy Spirit making a union with the human spirit was to govern the soul or mind and will. The spirit and mind together would use the body as the means for executing goodwill. (e.g. Proverbs 16:23: *The heart* (i.e. heart) *of the wise teacheth* (i.e. mind) *his mouth* (i.e. body), *and addeth learning to his lips* (i.e. body).)

Mental Capacity (Background)

Adam possessed the intelligence to name every living creature. This is the natural ability of man, created in the damah and tselem of God, and it demonstrates the greatness within us. Simply as God's creation, we are genius. Prior to the sin that separated humanity from God, we were spiritually, mentally, and intellectually a replica of God. As a replica, we possessed the ability to think, reason, and produce decisions that were in direct correlation with the design of God. The fact that we were not created to lack the free will of choice is evident in the fall of man. Even now it is important that we realize the capacity of brilliance within us.

We are a copy of the creative genius—God. In a single day He created an intelligent and intricately formed human being from the dust of the ground. The brain of that human is the most complex physical component in existence today. Even after Adam's decision to disobey God, the ingenuity, brilliance, and intelligence allotted man was further demonstrated in the descendants of Adam. The great-great-grandchildren of Cain, Adam's son, were tentmakers, musical instrument inventors, and instructors in brass and iron artifacts.[a]

The Brain—How It is Put Together

The brain has its own structure and function.[a] This complex organ weighs approximately three pounds. Dr. Daniel G. Amen, an American psychiatrist and brain-disorder specialist, describes the brain's appearance as a "mass of soft buttery tissue." This mass of tissue "uses twenty percent of the oxygen and blood pumped by the heart."[b]

Considering the brain's structure and its functions, we will look at some of the main components and their parts: the brainstem, cerebellum, thalamus, cerebrum, and cerebral cortex.

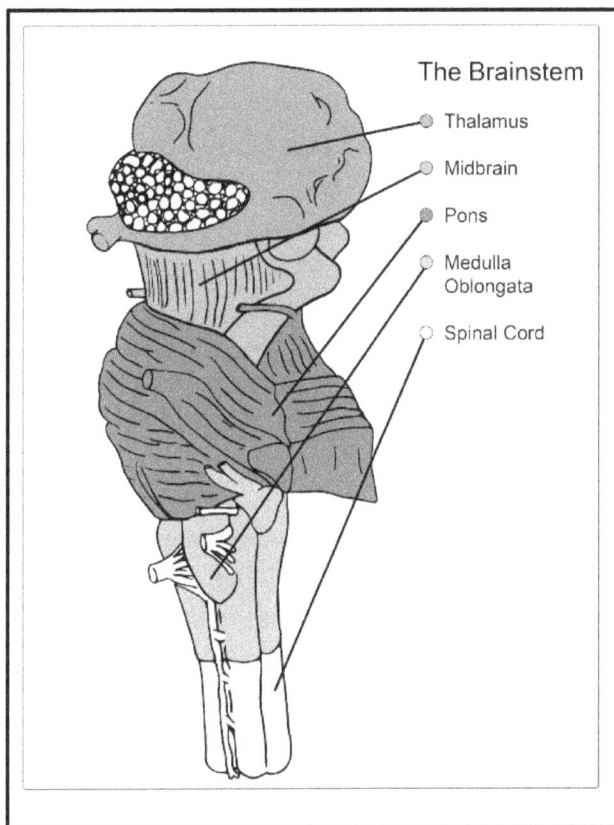

The Brainstem

- Thalamus
- Midbrain
- Pons
- Medulla Oblongata
- Spinal Cord

The brainstem is the connection between the brain and the spinal cord. It and its components function to provide support in regulating the cardiovascular and respiratory systems, consciousness, and sleep cycles. It is comprised of the *midbrain, pons,* and the *medulla oblongata.*

- The midbrain is the smallest region of the brainstem. It

functions as a relay station for the deciphering of auditory and visual systems. In conjunction with the basal ganglia, which select and mediate movement, the midbrain is associated with the movement of the eyes, body, and hearing.

- The pons, a white matter mostly made of nerve fiber, serves as a distributor for the brain and thalamus. The nerve fibers carry signals from the brain to the cerebellum and medulla and from the medulla to the thalamus. To be discussed later, the thalamus is the most prominent of all the components of the brain and their miscellaneous signals and connections.

- The medulla oblongata is responsible for maintaining vital body processes such as breathing and heart rate.

The overall responsibility of the brainstem is to assist in the maintenance of breathing, heart rate, and the movement of sensory information. Sensory information includes anything seen with the eyes, touched with a body part, tasted, smelled, heard, etc.

The cerebellum, known as the "little brain" because of its resemblance to the cerebral hemispheres, is located at the back of the brainstem where the spinal cord and brain join. It assists the midbrain and basal ganglia[4] with movement, and regulates posture, balance, coordination, and motor control. This enables fluid movement of the body and its various members.

This is accomplished by the close monitoring of sensory information received by the inner ear, nerve endings, and auditory system. The cerebellum is the arbitrator of motion memory—the ability of the body to repeat motions seamlessly and effortlessly. Although the cerebellum only accounts for approximately ten percent of the brain's volume, fifty percent of the brain's neurons are resident in this area.[c]

---

4 Basal Ganglia – A group of structures (Caudate Nucleus [body and tail] Globus Pallidus [exterior and interior], Subthalamic Nucleus, Substantia Nigra, and Putamen), coupled with the thalamus involved in the coordination of movement.

The thalamus has the responsibility of deciphering and relaying pertinent sensory signals received from each sense (touch, sight, hearing) except smell, to the cerebral cortex.[d]

It, along with the hypothalamus, are

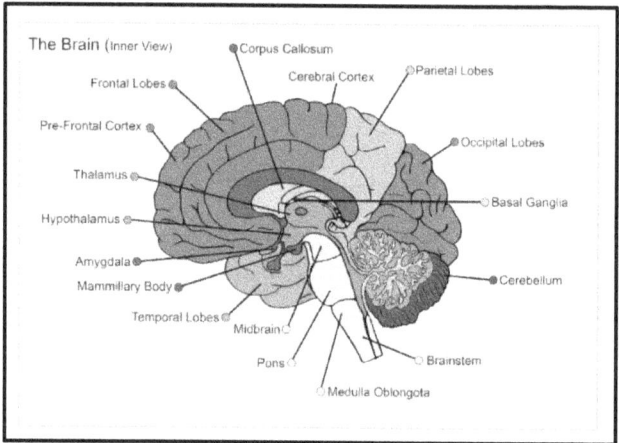

The Brain (Inner View)

Corpus Callosum
Cerebral Cortex
Parietal Lobes
Frontal Lobes
Pre-Frontal Cortex
Occipital Lobes
Thalamus
Basal Ganglia
Hypothalamus
Amygdala
Cerebellum
Mammillary Body
Temporal Lobes
Midbrain
Pons
Brainstem
Medulla Oblongota

portions of a division of the brain called the diencephalon.[5] These members, together with the specific sensory information received from the brainstem and spinal cord, form an elaborate relay system between the senses and the cerebral cortex. For example, visual information from the eye's retina is received, processed, and passed on to the visual cortex of the cerebral cortex. The information is either processed for spatial location—where things are

5 Diencephalon - the part of the brain that includes the basal ganglia, thalamus, hypothalamus, and associated areas. http://www.collinsdictionary.com/dictionary/english/diencephalon

in relation to the body—or visual form, such as facial recognition.

The hypothalamus controls the homeostasis of the body: temperature, heartbeat, and the circadian rhythm (the body's twenty-four-hour clock). It controls the balance of the body's internal environment.

The maintenance of the body's temperature is essential to health and wellbeing. The effect of an imbalance could cause convulsions if too warm and unconsciousness and brain damage if too cold. The effect of an imbalance in either the mental or physical body could be detrimental and possibly cause death.

The cerebrum is the last component that we will discuss. It occupies three-quarters of the brain's volume. The components associated with the cerebrum include the corpus callosum, lobes, and cerebral cortex. The surface of the cerebrum is covered by the brain's gray matter, known as the cerebral cortex. The cerebral cortex, consisting of eighty percent of the brain's mass, is divided into four sections, or lobes: *frontal*, *parietal*, *temporal*, and *occipital*.

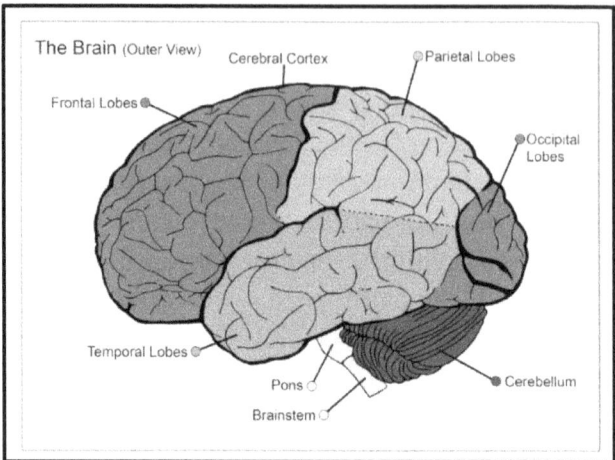

The Brain (Outer View)

The frontal lobe consists of the pre-frontal cortex, which is the brain's executor, and the motor cortex, which carries out the body movements planned by the prefrontal cortex.[6e] The pre-frontal cortex is where thinking, planning, focus, forethought, judgment, organization, impulse control, and empathy reside.

The *parietal lobe* handles sensation, reactions to the environment, directional sense and sensory processing.[f] Within the parietal lobe is the somatosensory cortex, where sensory information entering the brain via language and vision is managed.[g]

---

6 Prefrontal Cortex - the gray matter of the anterior part of the frontal lobe that is highly developed in humans and plays a role in the regulation of complex cognitive, emotional, and behavioral functioning.

The integration of signals from many of the body senses, including the physical location of body parts, occurs in the motor cortex of the parietal lobe. This area is also known as the dorsal route, or "where" pathway[h]—so called because it helps the brain know where things are located.

The temporal lobe, which resides on both sides of the brain in lateral positions, is the seat of auditory processing. In this area, things are recognized and named, and memories are placed into long-term storage. This is the location of the "what" pathway. In the "what" pathway, emotional reactions are regulated.

The amygdala located near the temporal lobe, along with the hippocampus, are proponents of emotional reactions. Emotional reactions are vital to long-term memory storage because of their parallel association with memory occurrence.

The primary function of the occipital lobe, at the back of the cortex, is vision.[i] Visual information from the world, gathered through the eyes, enters the occipital lobe for analyzation if the information requires locational data ("where"). The result is sent on to the parietal lobe and to the temporal lobe if recognition ("what") data is needed. The result yielded by this process is then forwarded to the frontal lobe for decision-making and execution.

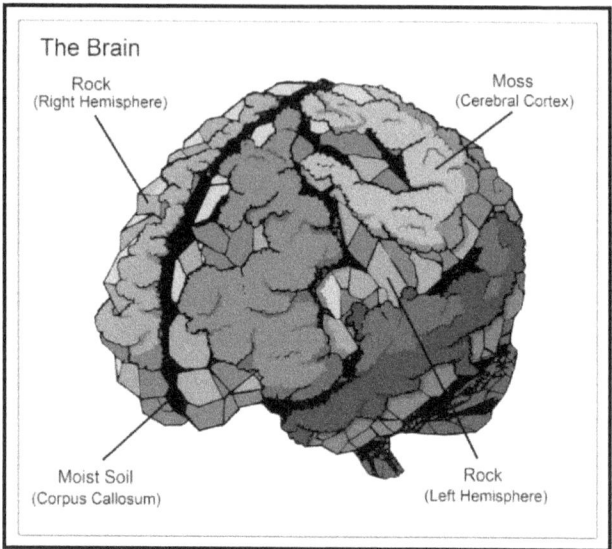

The Brain

Rock
(Right Hemisphere)

Moss
(Cerebral Cortex)

Moist Soil
(Corpus Callosum)

Rock
(Left Hemisphere)

The cerebrum is divided into two hemispheres: right and left. Between the hemispheres is the corpus callosum. The corpus callosum is made up of a band of nerve fibers which allows the hemispheres to constantly communicate with one another, like a middleman carrying messages between two parties. For a better concept of the make-up of these components, consider this for a visual understanding: two rocks (cerebrum) placed side by side with moist soil between them (corpus callosum) and covered by moss of four different shades of green (cerebral cortex).

The cerebrum, in association with the components of the frontal lobe, communes with every area of the mind and body. This, coupled with the emotional critique of the amygdala, sets the perfect stage for the making of decisions, memories, and morals.

In these areas of the brain exist the struggle between the recreated heart and the unregenerate mind. Here, where decisions become actions, memories are stored, and the consciousness of morals are established and executed, exists the perfect soil for *the God Seed*.

The Brain—How Its Parts Function

The brain is an electrochemical organ with voltage spikes and chemicals that travel through it and out into the body's nervous system.[a] For example, you are on your evening walk, and suddenly, you are startled by a dog barking from the top of a block wall that you are passing. This frightening occurrence will generate an electrochemical signal in the brain. This signal is known as an action potential[7] because it generates a significant chemical imbalance which will cause an action to be taken. So, in response to being startled, you begin to run or perhaps take an abrupt step away.

The full range of what takes place originates in the sensory division of the *peripheral nervous system* (PNS). The PNS receives information from the body's sensory receptors and transmits the input to the central nervous system (CNS) *which consists of the brain and spinal cord*. The CNS processes the

---

7 Action Potential—A momentary change in electrical potential on the surface of a neuron or muscle cell; nerve impulses are action potentials. They either stimulate a change in polarity in another neuron or cause a muscle cell to contract. The American Heritage© Science Dictionary

information to determine what must be the resulting response. When determined, the CNS returns the response to the motor division of the PNS, where it is delivered to the body to complete the resulting action.

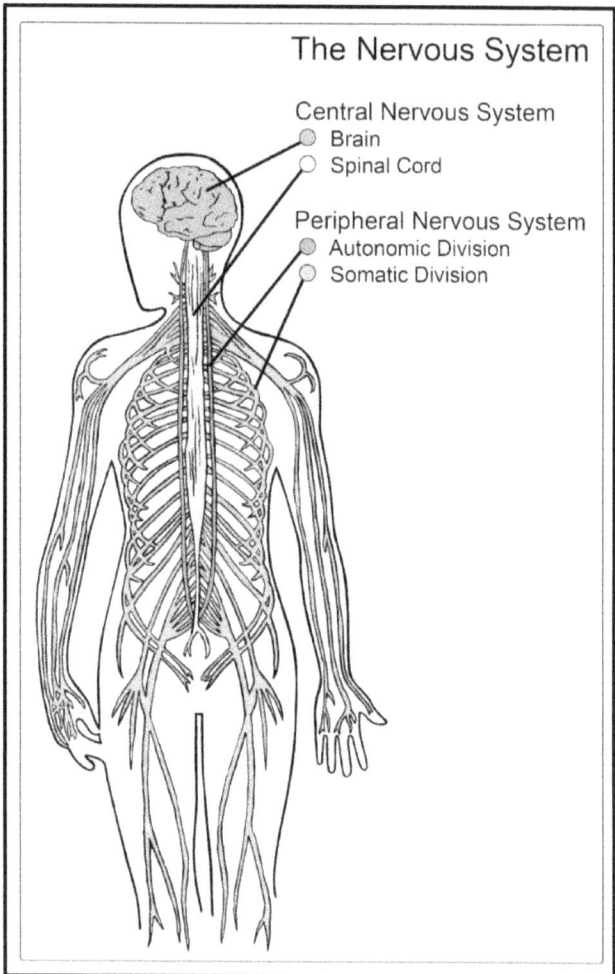

The Nervous System

Central Nervous System
- Brain
- Spinal Cord

Peripheral Nervous System
- Autonomic Division
- Somatic Division

The neuronal processes of the *dendrite, axon, and synapse* are utilized to carry out the response issued by the CNS. The dendrite is the neuronal receptor that transmits signals between neurons. The dendrite receives a signal from the synapse of the adjoining cell and transmits the signal on to the neuron to which it is associated. The neuron conducts the signal through its fibrous, vine-like extension called an axon.

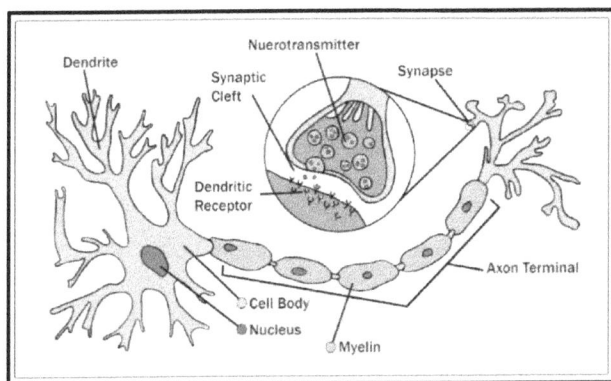

An axon is covered by a substance called myelin, which helps the axon to rapidly convey the received nerve impulse. The impulse travels along the axon to the synaptic terminal. At the synaptic terminal, chemicals called neurotransmitters are released. This chemical release initiates an action potential at the dendrite of the adjoining neuron body. The synapse makes the neuron-to-neuron connection. The neurotransmitter crosses the synaptic cleft—an immeasurable separation between neurons, invisible to the naked eye—to the receptors of the post-synaptic neuron. The signal continues through the dendritic receptor on to the adjoining cell to complete the intended purpose of the signal transmission. (i.e. fight or flight)

The voltage spikes of the brain are activated and fueled by the tightly balanced interplay of the neurotransmitters.[b] There are two categorical types of neurotransmitters: excitatory,[8] which encourages action, and inhibitory,[9] which hinders action. Some of the most common excitatory and inhibitory neurotransmitters are acetylcholine, dopamine, norepinephrine, and serotonin. Inhibitory neurotransmitters are composed of amino acids such as gamma aminobutyric acid (GABA), glutamate acid, aspartic acid, and glycine.

Acetylcholine, the first of the neurotransmitters, usually functions as an excitatory but can function as an inhibitory neurotransmitter. It was discovered in 1914 by Nobel Prize winner and neuropsychologist Henry Hallett Dale.[c] Acetylcholine acts as a neuromodulator[10] in both the central and

8 Excitatory—exhibiting, resulting from, related to, or producing excitement or excitation. (Merriam Webster, n.d.)
9 Inhibitory—to prohibit or prevent from doing something. (Merriam Webster, n.d.)
10 Neuromodulator—a neurotransmitter that is not reabsorbed by the pre-synaptic neuron or broken down into metabolite.

peripheral nervous systems. As a neuromodulator, instead of acting on individual neurons, it stimulates a variety of neurons throughout the nervous system. It plays a key role in learning, memory, and sending signals from motor nerves to muscles.[d]

Norepinephrine is primarily an excitatory neurotransmitter that promotes arousal or the "awake state" and focused attention. It is also considered a stress hormone affiliated with the "flight or fight" response.

Dopamine also has the dual-functioning capability of an excitatory and inhibitory neurotransmitter. It affects mostly motor arousal, memory, attention, and problem-solving.

Working in conjunction with dopamine and norepinephrine, Serotonin, an inhibitory neurotransmitter, helps to regulate arousal. It is also thought to be instrumental in the mood of an individual and the balancer for excitatory neurotransmitters. Serotonin must be chemically produced in the brain from tryptophan[11] because the serotonin found in the body's gastrointestinal tract cannot cross the blood-brain barrier. The blood-brain barrier is the brain's protective shield, made up of closely-

---

11 Tryptophan—an essential amino acid in humans. Can only be obtained from diet. Tryptophan is used to formulate serotonin and the hormone melatonin, used in control of the wake/sleep cycle.

knit blood vessels covered by endothelial[12] cells. These cells create a non-communitive barrier between circulating blood and the brain.

According to Shannon Moffett, author of *The Three-Pound Enigma*, "the brain is immunologically privileged, meaning the normal weapons utilized by the body to fight infection, such as medicines, cannot easily gain access."[e] A writing companion of mine called this "the holy of holies of the body."[13]

GABA is the major inhibitory neurotransmitter of the CNS, occurring in thirty to forty percent of all synapses. "GABA contributes to motor control, vision and other cortical functions."[f] The greatest concentration of GABA is in portions of the basal ganglia, hypothalamus, periaqueductal[14] gray matter, and the hippocampus.

The presence of GABA in the brain is two hundred to one thousand times greater than

---

12 Endothelial Cells—a thin layer of scale-like cells that line vessel walls and form an interface between circulating blood and the vessel wall.

13 The temple described in the Bible consisted of three parts: the outer court, inner court, and the holy of holies. Only select individuals were allowed into the holy-of-holies area of the temple.

14 Periaqueductal—of, relating to, or being the gray matter which surrounds the aqueduct of Sylvius (a channel connecting the third and fourth ventricles of the brain—also called *cerebral aqueduct and sylvian aqueduct*. (Merriam Webster, n.d.)

that of other neurotransmitters. The release of GABA into the synapse causes inactivity.[8] This inactivity reduces the action potential of the neuron to which it has bonded. GABA reduces the neuron's receptivity to subsequent electrical signals.

> *Footprint: This is a prototype of the function of the Holy Spirit, to prohibit one from aborting their God-ordained and -given purpose.*

GABA can be considered the red light of the CNS. Without it the neurons of the CNS would overload, never resting or resetting. As an example, it is speculated that such an overload, or lack of GABA, is possibly the cause of seizures.

The amino acid glutamate, the antagonist of GABA, is associated with learning and memory. It is the major excitatory neurotransmitter. When glutamate is bound to a glutamate protein receptor, the functions of this neurotransmitter are opposite that of GABA and, thereby, create a balance for GABA.

Aspartate is one more of the excitatory neurotransmitters that have the job of increasing the *likelihood of neuronal depolarization*[15] in the post-synaptic membrane.

---

15 Depolarization—loss of polarization; *especially,* loss of the difference in charge between the inside and outside of the plasma membrane of a muscle or nerve cell due to a change in permeability and migration of

The depolarization of a neuron causes the neuron to pass on action potential, preparing it for future functions. The hyperpolarization[16] of a neuron inhibits additional action potential of the neuron. The hyperpolarization of the membrane makes it less likely to depolarize or abort its purposed function.

Glycine, the last of the amino acids of the GABA group, is an inhibitory neurotransmitter. Acting along with aspartate, glycine also has a preparatory responsibility. The glycine neurotransmitter binds to a neuron receptor and causes the post-synaptic membrane to become more permeable,[17] or porous, for the acceptance of neurochemicals.

All neurotransmitters must have a point in which the transmitting of signals terminates. As in any process, if there is no end, a point of rest, or reset, there runs the danger of system overload and failure. There must be a maintained balance in the brain anatomy. For example, if serotonin, known to assist in the regulating of appetite, is out of balance, it could be the cause of one being a glutton, bulimic, or

---

sodium ions to the interior. (Merriam-Webster, Merriam-Webster, 2015)

16 Hyperpolarize—to produce an increase in potential difference across (a biological membrane). (Merriam-Webster, Merriam-Webster, 2015)

17 Permeable—having pores or openings that permit liquids or gases to pass through. (Merriam Webster, n.d.)

anorexic.

Some neurotransmitters, such as serotonin or norepinephrine are reset or cleared from the synapse by a process called reuptake. Reuptake is the reabsorption of the excess of a neurotransmitter chemical on the presynaptic neuron side. After the neural impulse signal has been transmitted, the chemical residue is absorbed by the pre-synapse receptors.

During this process of reuptake, the neuron returns to its neutral polarity, where the inside of the neuron has a negative charge created by potassium ions and the outside has a positive charge created by sodium ions. This negative/positive charge creates a vacuum-like suction comparable to that of a hole in an airborne plane fuselage.

An additional vital component of the brain's environmental health is the glial cell. These cells provide the necessary support for the proper functioning of the neuronal processes. This maintenance comes from various glial cells, most of which reside in the CNS:[h]

- The astrocyte glial helps to regulate blood flow to the brain, maintain the fluid composition surrounding neurons, and communication between neurons and the synapse.

- The microglia assist the immune system and serve as scavengers, removing dead cells, invading microbes, and clearing the debris of degenerating neurons.
- The oligodendrocyte and Schwann cells work together to create myelin.
- Satellite cells, thought to serve as a protective barrier, cover the cell bodies of the PNS.
- Ependymal cells line the vesicles of the brain and central spinal cord, promoting the circulation of the cerebrospinal fluid inside the ventricles of the brain and the spinal canal.

The glial cells also assist neurons during development to find their destination and contribute to the formation of the blood-brain barrier.

An understanding of the functions of the brain lays the foundation for comprehension of consciousness. Consciousness can be defined as one's ability to be aware of one's environment, culture, surroundings, thoughts, and feelings. The outcome of the combined stimuli, mixed with knowledge and experience, produces one's conscience.

According to Michio Kaku, in his book *The Future of the Mind*, the human brain operates in a level-three consciousness.[i] From a physicist's perspective, Kaku defines this level of consciousness as a "form of consciousness that creates a model of the world and then simulates it in time, by evaluating the past to simulate the future."[j] This means the brain has the ability to perceive possible future events or outcomes. This is an ability that only the human mind has the capability of performing and the capacity to maintain. He alludes that the mind arrives at decisions based upon current and previous experiences that have been stored and retrieved by the mind.

To decide or achieve a goal requires the ability to analyze in-coming sensory perception to formulate learned decisions. A learned decision uses programmed information that influences the results of a choice. An example of this concept is demonstrated in Ecclesiastes 12:1, *Remember now thy Creator in the days of thy youth, while the evil days come not, nor the years draw nigh....* In this verse, we are admonished *to remember* our Creator before the years of sin have taken their toll on our lives. The charge *to remember* implies something previously known.

Merriam Webster defines "remember" as the ability to bring an image or idea from the past into the mind. This Scripture suggests that there is a memory of the Creator that can be recalled. Man, therefore, has the capability of recalling and remembering his Creator through the analysis of incoming sensory perception to the formulation of learned decisions.

Consciousness and Memory

Attention to the area of consciousness is essential in demonstrating the distinctive inclination of the heart toward God and its ability to follow Him. Dictionary.com defines consciousness as:

> The inner sense of what is right or wrong, in one's conduct or motives, impelling one toward right action: to follow the dictates of conscience. The complex of ethical and moral principles that control or inhibit the actions or thoughts of an individual.

Strong's Exhaustive Concordance, as well as Biblestudy.com and the Greek Lexicon, have a collective definition of the etymological origin of the word "consciousness" that conclude consciousness to mean the ability of the mind/soul to see clearly, without uncertainty, to distinguish from good, bad, moral, and immoral. It is one's ability to apply general principles, gained from experience, to form a moral judgment that is perceived to be correct.

Consciousness is not a product of the

brain alone, but also of the mind[18]. The brain and mind, in conjunction with a person's surroundings, assist in the cultivation of this awareness. It is a culmination of all we think, feel, and experience. This process of consciousness cultivation helps to form the moral concepts that shape the fibers of one's life.

Francis Crick and James Watson state that consciousness is considered a phenomenon and the greatest problem of science.[a] Susan Blackmore wrote an article titled *Consciousness, An Introduction, Vol. 1,* which states that "consciousness is our first-person view of the world, one that is private and unique in each individual."[b]

Stuart Hameroff, M.D., Director of the Center for Conscious Studies, in a magazine article for *Science and Nonduality (SAND)*,[c] defined consciousness as "our existence." Peter Russell[d] in a *SAND interview* titled *A Conscious Chaos*, in dialogue concerning consciousness, said:

> The actual experience we have— the forms that appear in consciousness—our thoughts, sensations, perceptions, and feelings—that's all determined

---

18 Mind – the conscious mental events and capabilities in an organism.

by what goes on in the brain. If we change the state of our brain, we change our experience. But the brain does not create the faculty of consciousness.

An analogy would be that of light in a film projector. Every film projector has a source of white light at the center. The light shines through the film and takes on various colors. Basically, the film filters the light and gives it shape and form, and then we see the film on the screen, and we see all these forms and stories unfolding, and we get engaged in it. In the beginning, it was just light. The light is already there, and what the film does is shape the light.

The light of consciousness is already there in everything and in all of us. What the brain does is shape that consciousness into all the different things we experience, and then we get caught up and excited by what is going on in our minds. The brain shapes the light of consciousness into experience. Matter influences experience, but it doesn't create the ability to have experience.

The state of the mind controls the potential of consciousness, the decisions and choices that will be made.

Consciousness is the creator of the experiences from which memories are formed. The outcome concerning consciousness versus memory is that, prior to an experience becoming a memory encoded into the brain, it is first consciousness.

Merriam-Webster defines consciousness as the state or fact of being conscious (the ability to perceive, apprehend or notice with a degree of controlled thought or observation) or awareness, especially of something within one's self; the state or fact of being aware of an external object.

In connection with memory, the conscious state of the brain is the link between man's innate desire for God and God's datum for communication.

In connection with memory, the conscious state of the brain is the link between man's innate desire for God and God's datum for communication.

Consciousness is the memory equalizer. It bridges the gap between the input stimuli and its product, or result. It is like the operand between two numbers, the focal point of the mind, the magnet for thought. The scientific reasoning behind the process of consciousness is to provide a means by which the enormous quantity of stimuli the senses encounter can be scrutinized and processed for relevance in the precise moment of occurrence. An article found in *Cognitive Science*, written by Christof Koch, had this to say about the development of consciousness:

> Complex organisms and brains can suffer from informational overload. In primates, about one million fibers leave from each eye, carrying approximately one mega-byte per second of raw information. One way to deal with this deluge is to select a small fraction and process this reduced input in real time, while the non-attended portion of the input is processed at a reduced bandwidth. In this view, attention (the relay to consciousness) selects information of current relevance to the organism, while the non-attended data are neglected.[e]

The information selected then becomes a part of conscious thought and is given attention. The process by which attention is obtained is monumental.

There are two types of attention allocations: top down and bottom up. Top-down attention, also called endogenous or sustained attention, is voluntary. This type of attention occurs when one chooses to allow focus on a given object, sound, reaction, or region of space, etc. Bottom-up attention, called exogenous or transient attention, is involuntary. In short, this type of attention causes an interruption in the process of top-down attention. For example, a mother is in conversation with a friend when, suddenly, there is a loud thump from an adjoining room where her child is playing. Immediately, her attention diverts from the conversation to the sound from the adjoining room. Her attention diverts from top-down to bottom-up. Also, note that the time it takes for top-down memory to process is more extensive than that of its counterpart. Top-down attention is a more thought-intensive state of consciousness.

As noted by Rita Carter in an article titled *Top-Down, Bottom-Up*, there are three types of conscious awareness:

1. Awareness in the moment— the brain registers and reacts to moment-by-moment events but does not encode in memory; (bottom-up)

2. Conscious awareness—events are registered and encoded in memory; (top-down)
3. Self-consciousness—event registered and remembered. The person is conscious of doing so (top down and bottom up).

Consciousness is the product of data found to be relevant at the moment. The determination of the data relevancy is based upon something known and/or experienced. In the case of *the God Seed*, consciousness is based upon the previous knowledge and/or experience of man with his Creator during the time of Adam.

Scientific studies show that events from past generations genetically mark coming descendants for generations into the future. Spiritually, the same is true. Everything about man's existence can be traced back to this root of the Garden.

To show the dependence that knowledge and/or experience has upon one's decisions, consider this example: when there is something perceived to be a threat or dangerous, self-preservation is immediate, and it is only after the duress has passed that the action taken is examined. The response will be based upon the individual and their previous knowledge and/or experience with the specific threat or danger. In this scenario of bottom-up consciousness, the result is the effect of both or either knowledge or experience. The answer to the incoming sensory data will differ for everyone.

If we look again at the example of the mother in conversation and change the relationship of mother to simply another woman in conversation, the thump from the adjoining room may yield a different attention response. Consciousness is an experience/knowledge-based phenomenon that is the precursor of memory.

The function of consciousness, as defined by the physicist Kaku, when simulated by the unregenerate mind, will yield a different result than that of a mind led by a heart yielded to the Spirit of God. The brain performs millions of computations, yet only those that spark emotional firing of the sensory system are brought to the forefront of the mind for conscious attention. Much of the information processed by the brain is performed unconsciously. There are thoughts that roam our subconscious that never surface into consciousness and, therefore, are never consciously committed to memory yet are retained. A good example of such an unconsciously retained memory might explain what is known as *déjà vu.*[19]

---

19 *Déjà vu*—The illusion of having already experienced something being experienced (physically) for the first time. An impression of having seen or experienced something before. Dull familiarity. (The American Heritage College Dictionary, 4th Edition, 2002)

The Process of Memory Creation

A memory, as defined by Webster, is the power or process of reproducing or recalling what has been learned. Endel Tulving is an experimental psychologist and cognitive neuroscientist whose research on human memory has influenced psychological scientists, neuroscientists, and clinicians.[a] He defines memory as "the result of the preservation of the past in the organism that remembers." Memory, by all indications, is a recollection of information from experience, knowledge, or perception.

The renewal process of the mind, as referenced in Romans 12:2,[20] is the recalling of the native precepts of God, which are written in the annals of the heart. These are principles of memory that are existing and are part of the created design of the mind. This is the point of connection for the regenerated heart to communicate with the un-renewed mind. The scientific name for this type of memory is *implicit*—the process whereby unconsciously known information is stored.

"The network of some 10,000 miles of

---

20 Romans 12:2 *And be not conformed to this world: but be ye transformed by the renewing of your mind, that ye may prove what is that good and acceptable and perfect will of God.*

nerve fibers, called white matter, connects the various components of the mind, giving rise to everything we think, feel and perceive," states Carl Zimmer in the *National Geographic* magazine article, "Secrets of the Brain."[b] As God desires all things to have order, He has Himself exhibited this order in the design of the brain functions. Zimmer also comments that "the circuits of the brain intersect at right angles, like the lines on a sheet of graph paper." The journey of the neurotransmitter along this circuit[21] that bridges the synaptic cleft is short. The memory process begins when the fluid-filled vesicles[22] of the presynaptic terminal are released and received by the receptors of the postsynaptic terminal.[c]

The varied chemicals of the brain that assist in the processing of memory bolster varied responsibilities. The major memory chemicals

---

21 In the 1930s Otto Loewi, Henry Dale, and Wilheim Feldberg established that the signal which bridges the synaptic cleft—located between the presynaptic and postsynaptic cells—is usually a neurotransmitter that releases between the two terminals, diffuses across the synaptic cleft, and binds to receptors on the postsynaptic target cell.

22 Vesicle, a small, secretory vesicle that contains a neurotransmitter, is found inside an axon near the presynaptic membrane and releases its contents into the synaptic cleft after fusing with the membrane. "Synaptic Vesicle." *Merriam-Webster. com*. Merriam-Webster, n.d. Web. 5 Mar. 2016.

are dopamine, acetylcholine, GABA, serotonin, and glutamate. The chief brain component for memory is the limbic system. It is comprised of the amygdala, hypothalamus, thalamus, mammillary bodies,[23] and the cerebral cortex. The cerebral cortex, including the frontal, parietal, temporal, and occipital lobes, is located adjacent to the cerebral hemispheres on the interior side of the hippocampus. These chemicals and brain elements combine to process, create, and encode memories.

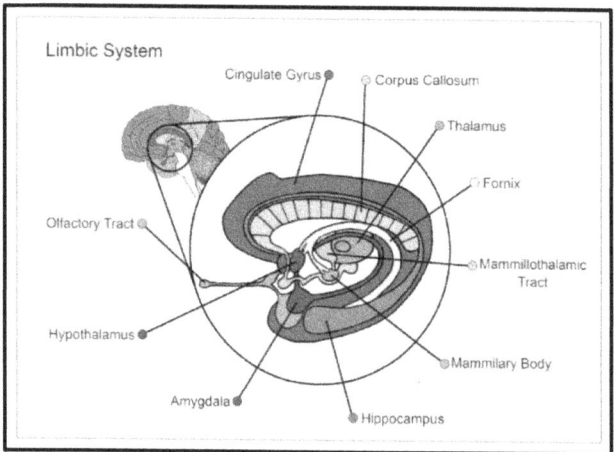

Limbic System

Cingulate Gyrus • Corpus Callosum • Thalamus • Fornix • Olfactory Tract • Mammillothalamic Tract • Hypothalamus • Mammillary Body • Amygdala • Hippocampus

There is a two-way communication system between the limbic system and the cerebral cortex. Information from the limbic system is fed to the frontal lobe to produce conscious feelings, while conscious knowledge

23 Mammillary bodies—the small limbic-system nuclei that are concerned with emotion and memory.

about the environment is taken from the cerebral cortex back to the limbic system in a continuous loop. This informational loop also receives sensory signals via neurons in the surrounding brain components. The type of electrical transmission communicated will determine which portions of the brain will react in response to the signals.

The process of memory retrieval is intricate and autonomous. For a memory to be retrieved, it first must have been stored. For a memory to qualify for storage, there must be an "experience giving rise to unusually prolonged and/or intense neural activity"[d] or emotion.

This causes the brain to encode the event as a memory. The encoding of memory is a process that engages the synchronous firing of the neuronal process. The progression that follows the engaging of the neuronal process depends on the type of memory being stored. For instance, if an experience involves the eyes, ears, and emotions, this pattern follows:

- the visual portion of the brain engages the temporal lobe, thalamus, and occipital lobe;
- the ears engage the brainstem and thalamus; while
- emotion engages the amygdala.

The reassembly of the experience must have the capability of retracing the process by which it was stored. This accumulation of

> *Footprint:*
> *Relationship is*
> *God's design for*
> *man.*

the memory is possible because neurons, via their synaptic connections, build a relationship with one another.

In an article published in the magazine *Science,*[e] it was noted that Santiago Ramon y Cajal found in his research that "neurons form highly scientific connections with one another, and these connections are invariant and defining for each species." Each

> *Santiago Ramon Y Cajal was a neuroscientist and physiologist whose research on the structure of the brain earned him the 1906 Nobel Prize in Physiology of Medicine.*

time a pathway is traveled to reproduce a stored memory, the relationship between the neurons is strengthened. Each time a memory is recalled, the potential for subsequent recall increases. For example, when studying for an exam, and not cramming, the constant revisiting of the material gives the studier a greater potential for recalling the material far beyond testing.

As outlined by Newberg and Waldman in *How God Changes Your Brain*, the process of memory creation or encoding is as follows:

1. Perception and attention to external stimuli are received in the pre-frontal cortex, causing two neurons to fire. This triggers a chain reaction in which one neuron, firing 5 to 50 times per second, can connect to up to 10,000 other neurons via 1,000-trillion synaptic connections.

2. From the synapse through the adjoining neuron and axon, electrical impulses, known as brain waves, are sent from one neuron to another by neurotransmitter chemicals distributed from the pre-synapse across the synaptic cleft to the post-synaptic receptor.

3. Sensory information is received by the thalamus and frontal lobe where the attention receives direction.

4. The amygdala is then aroused by emotions. The arousal of emotion causes the brain to respond to the received stimuli. This response is balanced by the hormone

oxytocin, which functions like an inhibitor, or calmer.

5. The produced response of the amygdala is then sent to the cerebral cortex for further processing. The greatest number of neurons are in the cerebral cortex. Here the received information is encoded. It is converted into a format for efficient processing.

6. The arranged information moves on to the hippocampus, where it is combined into a single experience and analyzed for the proper type of memory storage.

7. The experience is then stored as short- or long-term memory.[f]

All of this occurs in a matter of nanoseconds. The longest period for memory storage comes at the final stage of processing when an event is stored as long-term memory. Because memories are stored in multiple locations of the brain, this process can take minutes or days.

Memories are allocated to specific locations

based upon the physical component(s) involved in their creation. This process can be likened to a computer hard drive prior to the current solid-state technology. In previous technology,

> *Neuroscientists at the University of California, Riverside, report in the* Journal of Neuroscience *that they now may have an answer to this question. Their study provides for the first time a mechanistic explanation for how deep sleep (also called slow-wave sleep) may be promoting the consolidation of recent memories.*
>
> *During sleep, human and animal brains are primarily decoupled from sensory input. Nevertheless, the brain remains highly active, showing electrical activity in the form of sharp-wave ripples in the hippocampus (a small region of the brain that forms part of the limbic system) and large-amplitude, slow oscillations in the cortex (the outer layer of the cerebrum), reflecting alternating periods of active and silent states of cortical neurons during deep sleep. Traces of episodic memory acquired during wakefulness, and initially stored in the hippocampus, are progressively transferred to the cortex as long-term memory during sleep.*

information was stored in sectors, meaning in multiple locations. After a period, the data could become fragmented, or separated from its associated parts. When defragmented, the process entailed scanning the drive to locate the pieces of the various data to put them together again in a more efficient order on the drive. This was done to increase the speed of locating information on the drive. In the human brain, when this information is gathered from its varied storage locations and recompiled, it is a memory recollection.

Rita Carter in *The Human Brain Book* identifies five types of memories and ten memory storage locations. The ten memory storage locations are:

- Parietal lobe, spatial memory;
- Caudate nucleus,[24] instructive memory;
- Mammillary body, episodic memory;
- Frontal lobe, working memory;
- Thalamus, directs memory attention;
- Cerebellum, conditioned memory, events linked by time;
- Hippocampus, experience-based memory;

---

24 Caudate Nucleus—one of four parts of the basal ganglia existing within each cerebral hemisphere. The other three are the putamen, the globus pallidus, and the amygdala. One of the four basal ganglia in each cerebral hemisphere that comprises the larger and external nucleus of the corpus striatum, including the outer reddish putamen and two inner, pale-yellow globular masses constituting the globus pallidus—called also lenticular nucleus.

- Amygdala, emotional memory;
- Temporal lobe, general knowledge memory;
- Putamen,[25] procedural skill memory.[g]

The five types of memories are:

- Episodic, which provides a reconstruction of past experiences, including sensations and emotions;
- Semantic memory is non-personal, factual knowledge that stands alone;
- Working memory is the capacity to hold information in the mind for just long enough to use it, such as the dialing of a phone number;
- Procedural "body" memory consists of learned actions, such as walking and swimming; and
- Implicit memories are those we do not know we have.[h] Those which have been stored without the benefit of consciousness.

---

25 Putamen—as a part of the basal ganglia, it assists in the movement of limbs and learning.

Procedural memory is a subdivision of implicit memory and depends upon it to perform daily functions—those that are performed without the use of conscious thought or the recall of short- or long-term memory.

Implicit memory has another alter-component: explicit memory. Explicit memory utilizes a conscious effort to recall previous experiences that have been stored into memory.

In 1982 Jacoby and Witherspoon conducted a landmark study to determine the function of explicit and implicit memory. They first tested the explicit memory of a group of subjects by showing them a list of words and then quizzing their recognition of those words.

A second test of implicit memory was given to the same subjects in which words were flashed on a screen, many of which, unbeknownst to the participants, were presented in the previous list. The words which were a part of both tests were more readily recognized. The second test of the implicit memory held true even if, in the first test of explicit memory, the words were not recognized.

The passing of time does not affect the performance of implicit memory. A quote from a Livescience.com article titled *Implicit Memory* makes a comparable statement: "Implicit memory uses past experiences to remember things without thinking about them. The performance of implicit memory is enabled by previous experiences, no matter how long ago those experiences occurred."

In implicit memory resides the memory of man's first sinless state of mind and his need to seek the One with whom he once shared a very intimate relationship—the One Who loved him so deeply that the residue of their relationship yet resides within him. It is tacit that, without God and the Spirit of God, man is unable to tap into this memory. The access to this memory cannot be accomplished without another encounter with God, an encounter so intense that it invokes neuronal activity able to recompile the reminiscence of the only Lover of his soul, God.

It is the awakening of the neurons in this area by God's Spirit that permits this memory to be recalled. The Lord in His everlasting love continues to draw man unto Himself. Jeremiah 31:5 bears spiritual connotation to this restoration of man back to God:

> I've never quit loving you and never will. Expect love, love, and more love! And so now I'll start over with you and build you up again ... you'll resume your singing, grabbing tambourines, and joining the dance. You'll go back to your old work of planting vineyards on the Samaritan

hillsides, and sit back and enjoy the fruit–oh, how you'll enjoy those harvests![i]

Paraphrased, this scripture says: I will never stop loving you and you can always expect My love for you to remain. Although we have been separated by your choices, I wait for you to join Me on the dance floor, where we will together resume our marriage dance. You will resume your music and your song of love, joy, and, peace. You will return to your intended place with Me, where you shall reap a harvest forever.

What an emotionally, heartfelt experience this word from God portrays. The Lord is a groom anticipating the arrival of His bride—you and me.

Emotions play a large part in what comes to the forefront of the mind. Emotions are appealed to through the senses and are the root motivator of decisions, behavior, and actions. For example, when the serpent approached Eve in the Garden to lure her into enticement, he appealed to her senses and emotions. *And when the woman <u>saw</u>* (sense) *that the tree was good for food, and that it was pleasant to the eyes, and a tree to be <u>desired</u>* (emotion) *to make one wise, she took …* (Genesis 3: 6) Eve saw and desired.

In 1 John 2:16, we see the appeal to the emotions of the flesh in the form of lust. This scripture provides a clear parallel with the Genesis scripture: *For all that is in the world, the lust of the flesh* (good for food, author emphasis added)*, and the lust of the eyes* (pleasant to the eyes, author emphasis added)*, and the pride of life, is not of the Father, but is of the world.*

Senses, desires, and emotions are not intended to be portrayed as bad or evil; they are a divine part of our makeup. They are tools used to guide our passions into His divinely planned destiny.

Jeremiah 29:11 expounds: *For I know the thoughts that I think toward you, saith the Lord, thoughts of peace, and not of evil, to give you an expected end.*

Other scripture passages speak of God's thoughts and plans for us: declaring our end from our beginning (Isaiah 46:10); desiring only good for us (3 John 1:2); loving us with an everlasting love and drawing us in loving kindness (Jeremiah 31:3); etc. In addition to a renewed relationship with Him, He wishes to invoke the memories that remind us of who we are through a revitalizing encounter with Him.

The Plan and Science

The previously provided information concerning the brain components, neurotransmitters, consciousness, and memory come together in this section to form a cohesive concept of the innate ability of man to recollect memory of his Creator and the competence He has given us to follow, to agree, and to obey—to love Him.

Let's clarify the biblical meaning of obey. The word generally carries a negative connotation in the sense that it conveys that one must be submissive or vulnerable. But the original meaning is to hear, agree, and submit to what is heard.[a]

God's original plan was for man to love Him with all his heart, soul, and mind. In creating man in His likeness and image, God created us with the ability to do so. Scripture teaches that what or whom one loves dictates one's behavior. When a child loves its parents, obedience is easy because the child does not want to disappoint the parents. The pattern of God is such that He has furnished the ability in us to accomplish anything He asks.

God showed His love toward us in that while we were yet in our sin, and odious to Him due to that sin, He gave His Son to die in our stead to return us to Himself—because of love.

John 3, verses 13, 16, and 18, depict the love God demonstrated in the giving of His Son, Jesus, and what that has provided for us. Jesus is called the Son of Man and the Son of God.

> *13 And no man hath ascended up to heaven, but he that came down from heaven, even the Son of man which is in heaven. 16 For God so loved the world, that he gave his only begotten Son, that whosoever believeth in him should not perish, but have everlasting life. 18 He that believeth on him is not condemned: but he that believeth not is condemned already, because he hath not believed in the name of the only begotten Son of God.*
>
> *John 3:13,16,18*

As the Son of Man, He is the heir of God's promises to man. Jesus was the last sacrifice[26] required for the ultimate forgiveness of sin, providing eternal life for mankind. As the Son of God, He is a portrait of the man God created—the perfect, sinless man. "He is the unique, beloved, only eternally begotten, Son of the Father, very God."[b] The sacrifice of Jesus on the cross, His death, His burial and resurrection, reestablished for us the privilege of direct access to the Father. When man encounters God's

---

26 For more information concerning the sin offering necessary before Christ, see Leviticus Chapters 4 and 5.

Word and allows that encounter to access the innate *God Seed* link within his mind, he becomes the man God planned, both naturally and spiritually.

Romans 10:8 states: *The word is nigh thee, even in thy mouth, and in thy heart: that is, the word of faith, which we preach.* The Word of God is accessible and thereby His ways. His Word establishes His ways.

Words have creative power in the lives of men. When God created the world with words, the ability to create with words did not cease with creation. Spoken words are yet creating. When the Word of God has priority in one's life, when spoken, It creates. The ultimate purpose of the Word is to bring, create, and restore life. All that God spoke during creation brought life. When He spoke the moon, stars, and sun into existence, they brought life to the skies. Imagine day without the sun and night without the moon and stars. In the mind, the fundamental undertaking of *the God Seed* is to bring life and for that life to exist in abundance. God, in His omniscience, infinite knowledge, wisdom, and understanding, provided in man's mind an eternal connection to Himself—*the God Seed*.

> *Footprint: The origin of the word "provide" comes from the Latin* providere, *meaning to look ahead, prepare, supply.*

Research completed by the late Eugene d'Aquki at the University of Pennsylvania suggests that the neurological architecture of the mind is naturally calibrated to have and embrace spiritual perceptions.[c] From the biblical perspective, considering the unchanging nature of God, in the brain there remains a perfect and undefiled spiritual connection or seed that is waiting and longing for a refreshing drop of God.

Newberg and Waldman, in their joint effort, *How God Changes Your Brain*, wrote:

> God. In America I cannot think of any other word that stirs up the imagination more. Even young children raised in nonreligious communities understand the concept of God, and when asked, will willingly draw you a picture— usually of the proverbial old man with the long hair and beard. As children grow into adults, their picture of God often evolves into abstract images of clouds, spirals, sunbursts, and even mirrors, as they attempt to integrate the properties of a reality they cannot see.[d]

There is within us an instinctive, inborn unconscious seeking of God. This pursuit of Him is often denied or attributed to a search of other things, desires, positions, etc. We are designed to desire one greater than ourselves. In fact, in the opinion of Newberg and Waldman:

> If you contemplate God long enough, something surprising happens in the brain. Neural functions begin to change. Different circuits become activated, while others become deactivated. New dendrites are formed, new synaptic connections are made, and the brain becomes more sensitive to subtle realms of experience. Perceptions alter, beliefs begin to change, and if God has meaning for you, then God becomes neurologically real.[e]

In relation to the importance of the mind's status, here are additional scriptures to consider:

Romans 12:2

Ephesians 4:24b

Philippians 4:8

This neurological transformation brings light to God's amazing correlation between the scripture and our physical makeup. This also gives reason to why the Bible instructs that the Word should dwell in us richly, and should be meditated upon day and night, never departing out of our mouths.[f] If given attention, listened to, kept before our eyes and in the midst of our hearts, the Word will bring life to all of our beings.[27]

The neural functions, activated circuits, new dendrite, and synaptic connections spoken of by Newberg and Waldman are not physically new but are connections of existing pathways.

These are neural paths previously traveled. Man, created on the sixth day, was given the ability to perform the requirements of the Word, God being an intricate part of his physical makeup.

The Bible takes a strong position concerning the thoughts and intents of the mind, this being demonstrated through scripture passages such as Proverbs 23:7a: *For as he thinks in his heart, so is he,* and James 1:8: *A double minded man is unstable in all his ways.*

The mind is the most powerful natural component given to us at creation. It is the motor of the body and the mirror of the soul. It can traverse the spectrum between righteousness, unrighteousness, good and evil. It is the navigator of our choices.

---

[27] Proverbs 4:20-22

The Heart and Mind

It has been verified that the heart and brain possess the ability to function independently of one another. After the brain has been medically declared dead, the heart, when supplied with oxygen, can continue to beat. Additionally, the consciousness of the mind has been seen to continue functioning up to thirty seconds after the heart has ceased supplying blood to the brain.

Although connected through the neuronal and nervous systems of the body, the brain can and does operate independently of the heart and vice versa. It has been determined that, during the first stages of a baby's life, the heart is one of the first organs to form.

This, being among the first organs to develop, is then also one of the oldest at birth. The heart has ample time to gain control over the yet-forming mind, which is the optimal order. However, due to the heart's spiritual disconnection and the mind's naturally aggressive strength to pursue its own agenda, the battle for control of the body and its actions are lost to the mind.

Watchman Nee, in his book *Release of the Spirit,* quotes T. Austin-Sparks to describe this transaction between the mind and heart:

> We must be careful that, in recognizing the fact that the soul

(mind) has been seduced, led captive, darkened, and poisoned with self-interest, we do not regard it as something to be annihilated and destroyed in this life.

Through Christ there is hope for the renewal of the mind; therefore, it is not to be annihilated or destroyed.

The mind represents the soul, will, and emotion (what we do), and the heart represents the spirit of man (who we are). So, during the time prior to the heart's acceptance of Christ, or regeneration, the mind is in full control. The mind, receiving a full supply of stimulation from the world around it, dictates the actions of the heart. In turn, this determines the actions and behavior of the individual and who they ultimately become.

The heart's condition is that spoken of in Jeremiah 17:9: —*deceitful above all things, desperately wicked and filled with abominations*. Nothing of God can ever proceed from it.

Proverbs 6:16-19 provides the characteristic makeup of such an individual:

1. A proud look,
2. A lying tongue.

3. Hands that shed innocent blood.
4. A heart that devises wicked thoughts and plans.
5. Feet that are eager in running to mischief.
6. A false witness that speaks lies.
7. One that pioneers animosity among the brethren.

> *A proud look [the spirit that makes one overestimate himself and underestimate others], a lying tongue, and hands that shed innocent blood, a heart that manufactures wicked thoughts and plans, feet that are swift in running to evil, a false witness who breathes out lies (even under oath), and he who sows discord among his brethren.*
>
> *Proverbs 6:17-19, Amplified Bible*

For there to be a reversal of these characteristics, the deeds of the mind must be mortified to the point that the heart gains control. God must have a means by which to communicate with the innate mind of man—that portion of the mind that contains the remnant of God. The key to this communication is the Word of God.

Innate, as defined by the Merriam-Webster Collegiate Dictionary, applies to characteristics or qualities that are part of one's inner essential nature, not acquired after birth, but pre-existing. So there remains in the mind an immutable point of access and sensitivity to and for the divine purposes of God. This pre-existing inner essential nature is *the God Seed*. The existence of this inward, integral nature continues to thrive through the exercise of the neuronal pathways of implicit/procedural memory. The access of *the God Seed* through these pathways will return us to our designed mental normality.

By nature, man has a misconception as to what is normal, and as the years have progressed, this idea of normal has increasingly drifted further away from the principles of the Bible. Normal has been defined by Webster and others[a] as:

- Conforming to a type, standard, or regular pattern,
- usual,
- natural, or
- expected.

It is a concept of man, which is different for every individual in the world. The determination of normal is defined or demarcated by each person in the same way consciousness is decided, that is through mind/ environment, associations, and interactions. The reality is that, in the world, there is no normal. The word normal does not appear in the King James version of the Bible, which is considered one of the most accurate translations from the original text. I do not believe that this omission is an oversight or a coincidence. There was no need for such a word. If there ever could be a specific definition, God is the epitome of normal.

This divagation from God is due to the sin nature of man. Through the influence of his mind and environment, man has determined this state of departure from God as normal. King David testified as to the inception of this aberration in Psalms 51:5: *I was brought forth in [a state of] iniquity; my mother was sinful who conceived me [and I too am sinful].*[b] This is not about the sexual act of his wedded parents to conceive him, but to the sin nature inherited from Adam and Eve.

This sin nature is further confirmed in Romans 7:14-23. With the emphasis on verses 14 and 20-23, Paul recognizes that the law is spiritual, but we are carnal, bound to sin and unable to do otherwise outside of a regenerate heart and renewed mind. Note, in this scripture soul equates to the mind, which is the resident location of the will. New nature parallels to the heart/spirit and members represent the flesh or body.

We know that the Law is spiritual; but I am a creature of the flesh [carnal, unspiritual], having been sold into slavery under [the control of] sin. Now if I do what I do not desire to do, it is no longer I doing it [it is not myself that acts], but the sin [principle] which dwells within me [fixed and operating in my soul].

So I find it to be a law (rule of action of my being) that when I want to do what is right and good, evil is ever present with me and I am subject to its insistent demands. For I endorse and delight in the Law of God in my inmost self [with my new nature]. But I discern in my bodily members [in the sensitive appetites and wills of the flesh] a different law (rule of action) at war against the law of my mind (my reason) and making me a prisoner to the law of sin that dwells in my bodily organs [in the sensitive appetites and wills of the flesh].[c]

Paul recognized that in himself, his inward man, his heart (new nature, spirit), was a desire to obey, but in his mind the ability to do so was difficult to find.

Due to its inherent nature, for the mind to be receptive to the regenerated heart, the mind must be renewed, returned to its originally created state. The fact that the Bible mentions the necessity of this process of renewal is evidence that it is possible for the mind to exist in a renewed state. The word *renew* denotes a previous state of reason or disposition. As the un-regenerate state is of the Adamic inheritance, the regenerate state is our inheritance provided through the shed blood of Christ, the second Adam. This inheritance seeks to fulfill God's desire for all to be restored. Ephesians 1:11-12 states:

> *In Him* (Christ) *we also were made [God's] heritage (portion) and we obtained an inheritance; for we had been foreordained (chosen and appointed beforehand) in accordance with His purpose, Who works out everything in agreement with the counsel and design of His [own] will, so that we who first hoped in Christ [who first put our confidence in Him] have been destined and appointed to live for the praise of His glory!*[d]

It has always been God's plan and desire for us to be in fellowship with Him and providing a means to return our inheritance is to His praise and glory. In an even more concise manner, Romans 8:14-17 affirms:

> *For as many as are led by the Spirit of God, they are the sons of God.*
>
> *For ye have not received the spirit of bondage again to fear; but ye have received the Spirit of adoption, whereby we cry, Abba, Father.*
>
> *The Spirit itself beareth witness with our spirit, that we are the children of God: And if children, then heirs; heirs of God, and joint-heirs with Christ; if so be that we suffer with him, that we may be also glorified together.*

God wants to set the bound free and, once freed by His Spirit, adopt them as children and heirs to share in the blessing of Christ.

As part of the Adamic inheritance and the cause of the unregenerate mind, the heart became and continues to be deceitfully wicked. Consequently, for the mind to be renewed, there must first be a transformation of the heart.

The current state of the heart is one of total depravity. To understand that the heart must first be transformed before the mind can be renewed, let us consider Romans 10:9–10:

> *That if thou shalt confess with thy mouth the Lord Jesus, and believe in thine heart that God raised him from the dead, thou shalt be saved. For with the heart man believeth unto righteousness; and with the mouth confession is made unto salvation.*

Notice that it is first with the heart that man must believe and then with the mouth that he speaks of his faith and salvation in God. For the mouth to speak, the mind must have a concept of the intent of the heart. This is the beginning of the mind's renewal and the plumule of *the God Seed*.

Until the mind is renewed from its state of deceitfulness, it is unable to truly love, perceive, understand, or do the will of God. The process of renewing the mind is where the war between the spirit (heart) and soul (mind) resides. "We must see how the soul has to be smitten a fatal blow by the death of Christ as to its self-strength and government."[e]

In the book *Matters of the Heart*, author Dr. Juanita Bynum makes this statement:

> God saves and converts your spirit, which is where your heart is. Your mind, on the other hand, resists transformation. If your mind is not transformed, then the miracle "heart" that God has

placed in you will never be able to fully manifest in your new lifestyle.[f]

Once the heart is transformed, its agenda must be to reclaim the mind. Joseph Conrad, a novelist, and the author of *Heart of Darkness,* said, "The mind of man is capable of anything—because everything is in it; all the past as well as the future."[g]

Everything that is needed for all and whatever life encounters has already been provided and exists within everyone. Choices in life are the determining factor between success and failure. One of the wonderful things about God is that even amidst failure, success is only a decision or choice away. For example, when life has temptation, there is a means for avoiding the desire to compromise:

> *No temptation [regardless of its source] has overtaken or enticed you that is not common to human experience [nor is any temptation unusual or beyond human resistance]; but God is faithful [to His word—He is compassionate and trustworthy], and He will not let you be tempted beyond your ability [to resist], but along with the temptation He [has in the past and is now and] will [always]*

provide the way out as well, so that you will be able to endure it [without yielding, and will over-come temptation with joy].[h]

There are no individual occurrences in our lives that have not taken place in the life of someone else and victory not achieved—the path of choice being that which produces victorious results. This is true in all of life; this is true in obedience to God.

The preceding portions of this writing have been to set the stage for the heart's acceptance of Christ and the mind's renewal. From this point forward, the focus shall be upon the process of renewing the mind once the heart has chosen obedience to Christ.

Renewing the Mind

Via the architectural wiring of the brain, at the conversion of the heart, the process of renewing the mind begins immediately to accept encounters with God. Per Newberg and Waldman in their dual-authored volume, *How God Changes Your Brain*, the following alterations to the brain's perceptions occur when the heart encounters and yields to God:

- The occipital-parietal circuitry identifies God as an object that exists in the world.
- The parietal-frontal circuitry establishes a relationship between the two objects known as 'you' and 'God.' It places God in space and allows one to experience God's presence.
- The greater the activity of this circuit, the smaller the boundaries between man and God become.
- The frontal lobe creates and integrates all of man's ideas about God—positive or negative—including logic

used to evaluate religious and spiritual beliefs. It predicts the future relationship to God and attempts to, intellectually, answer all the why, what, and where questions raised by spiritual issues.

- The thalamus gives emotional meaning to concepts of God and gives a holistic sense of the world. The thalamus appears to be the key organ that makes God feel objectively real.

- The amygdala, when overly stimulated, creates the emotional impression of a frightening, authoritative, and punitive God, and it suppresses the frontal lobe's ability to think logically about God.

- The striatum[28] inhibits activity in the amygdala, allowing one to feel safe in

28 Striatum—a striped mass of white and gray matter in the brain which controls movement and balance. (Collins English Dictionary. Copyright© Harper Collins Publishers) A component of the basal ganglia, comprised of the caudate and putamen.

the presence of God.

- The anterior cingulate (related to top-down, bottom-up information) allows one to experience God as loving and compassionate. It decreases religious anxiety, guilt, fear, and anger by suppressing the activity of the amygdala.[i]

> *Footprint: This progression of implicit memories has all the indications of God's innovation and intervention for the restoration of our lives unto Him.*

The heart or, as previously established, the spirit of man, is also the place where the Spirit of God takes residence when He is accepted as Savior and Lord. The endeavor of the spirit is to convert or renew the mind (soul) to its decision to follow God. To effectively renew the mind through access to *the God Seed*—the seed which is buried beneath years of the soul's self-indulgence and influences of the world's exterior stimuli. This is a lifelong process that requires unrestricted access to the neuronal pathway and memory trace of implicit memory.

To refresh, implicit memory is a memory created and stored as long term without one being consciously aware of its creation. Under normal conditions, for memory allocation of either short- or long-term storage, there must be a stimulation of the brain that annunciates conscious awareness. It is an anomaly for long-term memory to be stored without this consciousness process.

To retain their relational synaptic connections, the neuronal paths to this allocated space of unperceived memory have been exercised through the recalling of other memories considered implicit, such as driving a car, walking, or riding a bicycle. Although implicit memory recall does not require conscious effort, by observation of other types of memories considered implicit, it can be deduced that they were at some point encountered or learned.

Long-term memory is stored as an explicit memory when the pre-frontal cortex determines that a received sensory is worthy of attention. The sensory information is transferred via the synaptic functions to the thalamus for direction and to the frontal lobe for processing the sensory information as short-term memory (also called working memory). The emotions aroused from the processing of working memory are filtered through the oxytocin-balanced amygdala for the proper reaction. The information is then sent to the hippocampus to combine and analyze the multiple facets or components of the working memory into a single experience for proper encoding and long-term storage.

Unlike explicit memory storage, which requires conscious efforts of the pre-frontal cortex, implicit memory is encoded and stored by the cerebellum, putamen, caudate nucleus, and motor cortex,[29] all of which bypass the need for conscious attention.

I believe that implicit memory is the storage area in which God chose to leave a trace of Himself. Its definition implicates a pattern of God: memories that are not retrieved consciously but are activated as part of a particular skill or action. Memories in the form of an emotion linked to an event that has not been consciously roused.[j] That, coupled with the definition of innate—existing in, belonging to, or determined by factors present in an individual from birth, not acquired after birth[k]—thus defines innate implicit memory as a memory stored outside the scientific realms of memory creation and which is a product present at birth. This can only be a concept of God.

The purpose of the attachment of other activities to this area of memory storage is to cause the neuronal activity of the implicit memory storage trace[30] to be in constant

---

29 Motor Cortex—a component of the cerebral cortex which influences smooth movement of the face, neck, trunk, and upper and lower extremities.
30 Memory Trace—a transient or long-term change in the brain that represents something (such as an

activity. For memories to be readily recalled, the path of their recall or recollection must be strong. The consistent recall function of other memory stimulating activities along the path of implicit memory preserves access to *the God Seed*, maintaining the pathway's availability, strength, and accessibility.

Memories have the potential to be altered or corrupted each time they are retrieved and reconsolidated. This corruption or false memory is called confabulation or refabricating. This occurs when a memory is recalled and intermingled with current events of the moment. When reconsolidated or returned to storage, the memory bears the potential of having incorporated items from what was taking place during its last recall.

To prevent the corruption or alteration of the precious *God Seed*, the implicit memory allocated for the purposes of God is only recompiled at the precise moment of our heart's regeneration. Upon its reconsolidation for the return to storage in long-term memory, it carries with it the refabricated events of the moment of salvation. From this time forward, each time *the God Seed* trace of implicit memory is recollected, via the prompting of the heart, it carries back to memory storage another faith milestone of the converted heart and renewing

experience) stored as a memory. (Merriam-Webster.com/dictionary, 12/2/2016) mind. Romans 1:17 illustrates the process in these words:

> For in the Gospel a righteousness which God ascribes is revealed, both springing from faith and leading to faith [disclosed through the way of faith that arouses to more faith]. As it is written, the man who through faith is just and upright shall live and shall live by faith.[l]

This is what takes place when *the God Seed* is recalled, reconsolidated, and returned to memory storage—the faith brought forth from storage returns to storage with a greater revelation of faith in God, thereby increasing the mind's neuronal capacity to trust in the reality, truth, and existence of God. This expands the mind's capacity to submit to the will of the heart which has submitted to God. Proverbs 4:20-22 declares that when we attend unto the Word of God, listen to it and keep it in the midst of our hearts, it will be health to all of our flesh, which includes our minds.

Memory is easier to recollect when associated by context. With the God encounter of the heart, which is the trigger for the recollection of *the God Seed* trace of implicit memory, the association of the heart's divine connection and supremacy over the soul and body are again made conscious. This causes an "aha" moment in which the heart recalls and begins to activate its rightful authoritative position over the mind and body.

An epiphany of the Genesis creation occurs when our spirit recalls it was once one with its Creator, Whose breath gave it life. The spirit, now exerting authority over the soul and body, proclaims a carnal existence was never destined for us as a triune unit of man. We have an inheritance in Christ prepared through His shed blood on the cross. No longer will we accept the corrupted thoughts of an unregenerate mind to have dominion of us as a unit. Our response to the Spirit of God is yes, and we will completely yield to the congenital *God Seed* planted at our creation.

In view of all the mercies of God, we will make a collective decision to dedicate the body of our triune being as a living sacrifice, one holy, devoted, consecrated, and well-pleasing to God; this being the minimal responsibility. The mind of our being shall be continually transformed to accept the ideals of the Spirit of God.[m]

This is now the lifelong endeavor of the spirit of man: to be wholly acceptable to God— spirit, body, and soul.

Conclusion

The plan of God is perfect. He, in His awesome wisdom and knowledge, created from the dust a man equipped with everything he would possibly ever need for success. In the midst of creation, the Creator, whose aversion to sin can be compared to no other, planted *the God Seed* in the mind of His creation. This would provide a way of return when, by the creation's choice, sin would happen. By means of implicit memory storage, God impregnated the mind of man with the God-given, inbred potential to love, to follow, to agree, to obey, and to return to his Creator via *the God Seed*—if he chooses to do so.

With the heart now in full control of the mind through *the God Seed*, neuronal connections of the dendrites, axons, and synapses have strengthened relationships to one another. The paths through the various regions of the mind now produce thoughts of good and not evil. The thoughts of the renewing mind have become as Paul admonishes in Philippians 4:8:

> *… whatsoever things are true, whatsoever things are honest, whatsoever things are just, whatsoever things are pure, whatsoever things are lovely,*

whatsoever things are of good report; if there be any virtue and if there be any praise, think on these things.

Epilogue

Tree roots develop in size and depth to support the height and weight of the tree above. A foundation is designed to carry what it supports. The foundation of a structure carries the weight of itself and whatever is placed on it, including the forces applied to it, such as wind, rain, earthquake, etc.

God has laid for us a foundation in support of a life of harmony and peace. It has been further fortified by the shed blood of Christ, Who is the cornerstone.

In our life construction, we need only build upon the foundation already provided by following the prescriptive code of the Bible. This will guide us back into the heavenly places of God.

When we learn to live from the heavenly places from which we were created, life for us on Earth will be like that prepared for us in heaven—His kingdom will then come on Earth as it is in heaven. All of this is accessible through the activation of *the God Seed*. For within us lies the same power that raised Jesus from the dead. This power enables us to reactivate the authority given us as those with dominion over the earth.

Let this mind be in you which was also in Christ Jesus.[31]

With the mind of Christ, affecting change in our sphere of influence is inevitable. We will perpetuate a butterfly effect with the propensity to affect lives for generations to come.

---

[31] Philippians 2:5

God has made you for Himself, and your heart is restless until it has found its rest in Him.

Ravi Zacharias

Bibliography

About Education. (2014). Retrieved from
http://psychology.about.com/od/aindex
/g/acetycholind.htm

Amen, D. (n.d.). Neurology 101 The Science of
the Brain. *Christian Counseling Today*,
pp. 22-26.

Andersen, P. (2014, March 6). *YouTube*.
Retrieved May 2014, from Bozeman
Science: bozemanscience.com

Anderson, P. (2014, March 6). *Bozeman
Science*. Retrieved June 14, 2014, from
The Brain: Structure and Function:
https://www.youtube.com/watch?v=k
M kc8nf PATI

Bailey, R. (2015, July 21). *About.com*. Retrieved
from Anatomy of The Brain:
http://biology.about.com /anatomy/p/c
erebrum.htm

Best, B. (2014, September 14). Retrieved from
http://w ww. ben
best.com/science/anat mind/anatmd10
.html#contents

Cajal, R. (n.d.). Breaking Down Scientific
Barriers to the Study of the Brain and
Mind.

Campanile, C. (2012, November 23). Americans
Are Getting Fatter: Poll. *New York Post*.

Carter, R. (2009). A Journey Through The Brain. In R. Carter, *The Human Brain Book, an Illustrated Guide to its Function, Structure, and Disorders* (pp 15-35).

Carter, R. (2014). *The Human Brain Book.* new York: DK Publishing.

Cherry, K. (2015, July 21). *About Education - Anatomy of the Brain.* Retrieved from psycology. about.com: http://psycology.about.com/od/biopsyc ologyss/brainstructure.htm#step5

Hebrew-Greek Key Word Study Bible, King James Version. (1991). Chattanooga: AMG.

Hub, B. (2015, December 30). *Strong's Concordance.* Retrieved from Bible Hub: http:// biblehub.com/ greek/5219.htm

Kaku, M. (2014). The Future of The Mind - The Scientific Quest to Understand, Enhance, And Empower The Mind. USA.

Kandel, E. R., & Squire, L. R. (2000). Breaking Down the Scientific Barriers to the Study of the Mind and Brain. *Science*, p. 1113.

Koch, C. (2006). Attention and Consciousness: Two Distinct Processes. *Cognitive Science*, pp. 16-22.

Lundbeck Institute. (2011, December 20). Retrieved from CNS Forum: http://www.brainexplorer.org /neurolog ical_control/neurotransmitters.html

Mastin, L. (2010). *Memory Encoding*. Retrieved
from The Human Memory:
http://www.human-memory.net

medicineNet.com. (2012, March 19). Retrieved
July 1, 2014, from
http://www.medterms.com

Merriam Webster. (n.d.). Retrieved June 22,
2014, from m-w.com:
http://www.merriam-webster.com/dicti
onary/pia mater

Merriam-Webster. (2015, August 9). *Merriam-
Webster*. Retrieved from Merriam-
Webster: http://www. merriam-
webster. com/dictionary /hyperpolarize

Merriam-Webster. (2015, August 9). *Merriam-
Webster*. Retrieved from Merriam-
Webster: http://www. merriam-
webster. com/dictionary/depolarization

Moffett, S. (2006). *The Three-Pound Enigma*.
New York: Algonquin Books of Chapel
Hill.

Morris, H. M. (2012). *The Morris Henry Study
Bible.* Green Forest: Master Books.

Nee, W. (1965). Introduction. In W. Nee, *The
Release of The Spirit* (pp. 6-8). Sure
Foundation.

Newberg, A. W. (2006). *Why We Believe What
We Believe.* New York: Free Press.

Newberg, A., & Waldman, M. R. (2009). *How
God Changes Your Brain.* USA:
Ballantine Books.

Pinto, Y., & Leij, A. S. (2013). Bottom-Up and Top-Down are Independent. *Journal of Vision*.

Seung, S. (2012). *Connectome.* New York: Houghton Mifflin Harcourt.

Strong, J. (n.d.). Strong's Exhaustive Concordance of The Bible. *The Exhaustive Concordance of The Bible.* McLean, Virginia, USA: MacDonald Publishing Company.

Strong's Concordance. (2015, December 30). Retrieved from Bible Hub: http://biblehub.com/greek/521 9.htm

Tancredi, L. (2005). *Hardwired Behavior-What Neuroscience Reveals About Morality.* Cambridge: Cambridge University Press.

The Amplified Bible. (1987). Grand Rapids: Zondervan.

The Amplified Bible. (1987). Grand Rapids, MI: Zondervan.

The Brain From Top to Bottom. (2014, September 20). Retrieved from Neurotransmitters: http://thebrain.mcgil l.ca/flash/i/i_01/i_01_m/ i_01_m_ana/i_ 01_m_ana.html

Tipler, F. (2007). *The Physics of Christianity.* New York: Doubleday.

Youtube - zooming in on the human brain. (2013, June 8). Retrieved June 14, 2014,

from Allen Institute:
www.youtube.com/user/AllenInstitute

Zimmer, C. (2014, February). Secrets of the
Brain. *National Geographics*, pp. 35-60.

Endnotes

(For detailed information of sites in the endnotes, please see the bibliography.)

The Point of It All
[a] (Campanile, 2012)
[b] (2 Corinthians 4:3,4)
[c] Galatians 5:22,23

The Origin of Man
[a] Genesis 2:8-9a, 10a
[b] (The Amplified Bible, 1987) Matthew 6:33
[c] (The Amplified Bible, 1987) Galatians 5:22,23
[d] Strong's Exhaustive Concordance of the Bible, Hebrew 120
[e] (Strong)
[f] (Nee, 1965)
[g] Watchman Nee – A prolific writer, teacher, and church planter throughout China during the twentieth century. For the last twenty years of his life, Nee was persecuted and imprisoned for his faith in Christ.

Mental Capacity (Background)
[a] Genesis 4:20-22

The Brain Components—How it is Put Together

[a] (Andersen, 2014)

[b] (Amen)

[c] (About Education, 2014)

[d] (Andersen, 2014); (Carter, A Journey Through The Brain, 2009)

[e] (Amen)

[f] (Amen)

[g] (Andersen, 2014)

[h] (Carter, A Journey Through The Brain, 2009); (Andersen, 2014)

[i] (Amen)

The Brain Functions—How They Work Together

[a] (Zimmer, 2014)

[b] (Newberg A. W., 2006)

[c] (Best, 2014)

[d] (Andersen, 2014)

[e] (Moffett, 2006)

[f] (The Brain From Top to Bottom, 2014)

[g] (Best, 2014)

[h] https://www.khanacademy.org/science/biology/human-biology/neuron-nervous-system/ a/overview-of-neuron-structure-and-function, July 4, 2017

[i] (Kaku, 2014)

[j] (Kaku, 2014)

Consciousness and Memory

[a] Reference Here

[b] Reference Here

[c] (Hameroff, A Brief History of the Study of Consciousness, Science and Nonduality)

[d] Expert in experimental psychology; *https://www.peterrussell.com/pete.php*, May 18, 2018

[e] (Koch, 2006)

The Process of Memory Creation

[a] https://en.wikipedia.org/wiki/EndelTulving

[b] (Zimmer, 2014)

[c] (Seung, 2012)

[d] (Carter, The Human Brain Book, 2014)

[e] (Kandel & Squire, 2000)

[f] (Newberg & Waldman, 2009)

[g] (Carter, A Journey Through The Brain, 2009)

[h] (Carter, A Journey Through The Brain, 2009)

[i] (The Amplified Bible, 1987)

The Plan and Science

[a] (Strong's Concordance, 2015)

[b] (Morris, 2012)

[c] (Newberg A. W., 2006)

[d] (Newberg & Waldman, 2009)

[e] (Newberg & Waldman, 2009)
[f] Colossians 3:16; Joshua 1:8

The Heart and Mind

[a] (Merriam-Webster, Merriam-Webster, 2015)(Google.com, 2018) (dictionary.com, 2018) (oxford Dictionary.com, 2018)
[b] (The Amplified Bible, 1987)
[c] (The Amplified Bible, 1987)
[d] (The Amplified Bible, 1987)
[e] (Nee, 1965)
[f] (Bynum, Matters of the Heart, 2002)
[g] (Kaku, 2014)
[h] Bible Gateway, Amplified Bible, 1 Corinthians 10:13

Renewing the Mind

[i] (Newberg & Waldman, 2009)
[j] (Carter, The Human Brain Book, 2014)
[k] (Merriam Webster, n.d.)
[l] (The Amplified Bible, 1987)
[m] (The Amplified Bible, 1987)Romans 12:1,2

9780578706894